THE NEWER ALCHEMY

THE
NEWER ALCHEMY

Based on
The Henry Sidgwick Memorial Lecture
delivered at Newnham College
Cambridge
November 1936

BY

LORD RUTHERFORD, O.M., F.R.S.

*Cavendish Professor of Experimental Physics
in the University of Cambridge*

CAMBRIDGE
AT THE UNIVERSITY PRESS
1937

CAMBRIDGE
UNIVERSITY PRESS

University Printing House, Cambridge CB2 8BS, United Kingdom

Cambridge University Press is part of the University of Cambridge.

It furthers the University's mission by disseminating knowledge in the pursuit of education, learning and research at the highest international levels of excellence.

www.cambridge.org
Information on this title: www.cambridge.org/9781107440425

First published 1937
First paperback edition 2014

A catalogue record for this publication is available from the British Library

ISBN 978-1-107-44042-5 Paperback

PREFACE

This little book contains in a somewhat expanded form the subject matter of the Henry Sidgwick Memorial Lecture delivered at Newnham College on November 28, 1936.

Since the early days of Radioactivity, the problem of the transmutation of the elements has occupied much of my attention, and I have followed with the greatest of interest and enthusiasm the remarkable increase in our knowledge that has come so rapidly in the last few years. This advance has been largely due to the development of new and powerful methods of attack on this general problem. For this reason I have thought it of interest to add a brief account of the new apparatus and methods which are now in common use in many Laboratories throughout the world.

I have also included a selection of the illustrations which I gave in my lecture, and would like to express my thanks to Professor C. T. R. Wilson, Professor E. Lawrence, Professor P. M. S. Blackett, Professor J. Chadwick, Mr P. I. Dee, Dr C. W. Gilbert, Professor

H. J. Taylor and Dr M. Goldhaber for permission to reproduce some of their photographs. I am grateful also to Dr Oliphant and Mr P. I. Dee for their help in preparing the illustrations and in correcting the proofs.

RUTHERFORD

February 1937

LIST OF PLATES

vii

THE NEWER ALCHEMY

IN this lecture I shall give a brief account of modern work on the transmutation of the elements. The title is intended to suggest a contrast to that ancient form of alchemy which had such an extraordinary fascination for the human mind for nearly two thousand years. The belief in the possibility of the transmutation of matter arose early in the Christian Era. The search for the Philosopher's Stone to transmute one element into another, and particularly to produce gold and silver from the common metals, was unremittingly pursued in the Middle Ages. The existence of this idea through the centuries was in no small part due to a philosophic conception of the nature of matter which was based on the authority of Aristotle. On this view, all bodies were supposed to be formed of the same primordial substance, and the four elements, earth, air, fire and water, differed from one another only in possessing to different degrees the qualities of cold, wet, warm and dry. By adding or subtracting the degree of one or more of these qualities, the properties of the matter should be changed. To the alchemists, imbued with these conceptions, it appeared obvious that one substance could be transmuted into another if only the right method could be found. In the early days of Chemistry, when the nature of chemical combination was little understood, the marked alteration of the appearance and properties of substances by chemical action gave support to these views. From time to time there arose a succession of men who claimed to have discovered the great secret, but we have the best reasons for believing that not a scintilla of gold was ever produced. When we look back from the standpoint of our knowledge to-day, we see that transmutation was a hopeless quest with the very limited facilities then at the disposal of the experimenters. With the development of experimental science and the steady growth of chemical knowledge, the ideas of transmutation were gradually discarded and ceased to influence the main advance of knowledge. At the same time these old alchemistic ideas have persisted in the public mind, and even to this day impostors or deluded men appear who claim to have a recipe for making gold in quantity by transmutation. These charlatans are often so convincing in their scientific jargon that they disturb for a time the sleep of even our most hard-headed financiers. We shall see that it is now possible by modern methods to produce exceedingly minute quantities of gold, but only by the transmutation of an even more costly element, platinum.

As the knowledge of Chemistry grew the old idea of transmutation was seen to be untenable. It was found that matter could be resolved into eighty or more distinct elements, the atoms of which appeared to be permanent and indestructible. The ordinary physical and chemical forces then at our command appeared to be unable to alter in any way the atoms of the elements. This idea of the permanency of the atoms received a rude shock when it was found in 1902 that the atoms of two well-known elements, uranium and thorium, were undergoing a veritable process of spontaneous transformation, although at a very slow rate. This conclusion followed from the discovery of the radioactivity of these two heavy elements which spontaneously emit penetrating types of radiation capable of blackening photographic plates

3

and discharging an electrified body. This radio-activity is a sign of the instability of the atoms concerned. Occasionally an atom breaks up spontaneously with explosive violence hurling from it either a fast α- or β-particle. The α-particle is a charged atom of helium of mass 4 which is shot out at a speed of about 10,000 miles per second. The β-particle, which is another name for the light negative electron, is generally expelled with a much higher speed. Sometimes a penetrating radiation of the X-ray type, known as the γ-rays, accompanies the transformation.

RADIOACTIVE TRANSFORMATIONS

If we take a gram of the element uranium, about 24,000 atoms break up per second with the emission of an α-particle. Yet the number of atoms in a gram is so great, that it would take about 4500 million years before half of the atoms are transformed. As a result of the emission of an α-particle of mass 4 from the uranium atom of weight 238, a new atom is formed of atomic mass 234. The atoms of this new element are very unstable and break up rapidly with the emission of a swift β-particle from each atom.

This process of transformation, once started, continues through a succession of stages, each unstable atom giving rise to another. The well-known element radium has its origin from the transformation of uranium, and is the fifth product in the series.

The activity of a radioactive body, measured by the specific radiation it emits, diminishes with time according to a geometrical progression. If the activity falls to 1/2 in a time T, known as the half-period, it falls to 1/4 in a time $2T$, to 1/8 in a time $3T$, and so on. It can readily be calculated that in a time $20T$ the activity has decreased to less than one-millionth of the initial value. This law of decay holds universally for all radioactive bodies; but the half-period T, which has a characteristic value for each active body, varies enormously for different substances. For example, the half-period of uranium is 4500 million years and for radium 1600 years, but is only one-millionth of a second for one of the products of radium known as radium C'. This law of decay is an expression of the fact that the number of atoms breaking up in unit time is on the average always proportional to the number of unchanged atoms present. Such a result is to be expected if the individual atoms break up according to the laws of chance.

The wonderful sequence of transformations which occur in uranium is shown in Fig. 1, where the circles represent the nuclei of the successive atoms which are formed. The half-period of transformation is added below, while the nature of the particle expelled, whether α or β, is indicated. It would take too long to discuss the methods by which this sequence of changes has been definitely established, but attention should be drawn to the extraordinary simplicity of the relations which connect together the whole series of transformations.

We now know that the chemical properties of an element are defined by its atomic number, which also represents the number of natural units of charge on the atomic nucleus. Since electricity is atomic in character, the nuclear charge is always given by a whole number, which varies from 1 for the lightest nucleus, hydrogen, to 92 for the heaviest element, uranium. The atomic number of each nucleus and also its atomic mass in terms of $O = 16$ are shown within the circles

The α- or β-particle which is liberated has its

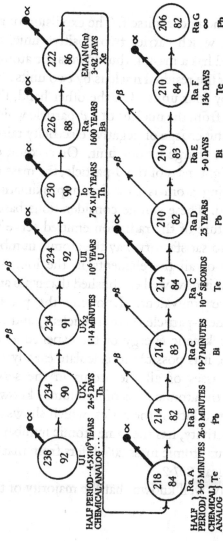

Fig. 1. Uranium series of elements. The upper number in each circle gives the atomic mass, the lower the atomic number and nuclear charge. The length of the broad arrow shows the relative distance of travel of the α-particles.

origin in the nucleus itself. The expulsion of an α-particle, which carries two positive units of charge and has a mass 4, thus lowers the atomic number of the residual nucleus by two units and its mass by four units. On the other hand, the expulsion from the nucleus of a β-particle which carries a unit charge of negative electricity raises the nuclear charge by one unit. On account of the very light mass of the β-particle, the mass of the resulting atom is to a first approximation unchanged. These simple considerations based on the nature of the radiation emitted serve to explain in a satisfactory way the atomic number and mass of all the elements in the long sequence. It is now well established that mass and energy are equivalent. Knowing the precise mass of the α-particle (helium nucleus) and the maximum kinetic energy of the expelled α- or β-particle, it is possible to calculate exactly the atomic masses of all the atoms in the series provided the atomic mass of uranium is known. The end product of the series, which shows no trace of activity, has the same atomic number as lead, but its atomic mass 206 differs from that of ordinary lead 207·2.

It is now well known that the majority of the

elements consist of a mixture of isotopes, i.e. of atoms which have the same nuclear charge but different masses. Aston has shown that ordinary lead is made up of at least three isotopes of atomic masses 206, 207 and 208, of which the mass 206 is the most abundant. The final product of the uranium series, generally known as uranium-lead, is thus one of the isotopes (206) of ordinary lead. The lead separated from an old uranium mineral mainly consists of this lead isotope 206. It will be noticed also that two radioactive isotopes of lead, atomic number 82, appear in the series, viz. radium B of mass 214 and radium D of mass 210.

It should be mentioned that a similar long sequence of transformations is shown in the elements thorium and actinium. The end product of thorium is again an isotope of lead, but of mass 208 instead of 206 as in the case of uranium-lead. The lead separated from a pure thorium mineral mainly consists of the 208 isotope. The end product of the actinium series of transformations is also an isotope of lead but of mass 207. It is a striking fact that the final product of the transformation of all three series is in each case an isotope of lead but of different atomic mass.

The remarkable changes in the chemical and physical properties of successive radioactive elements is well illustrated by the transformation of radium. In the pure state, the element radium is a metal with chemical properties resembling those of barium. It breaks up with the emission of α-particles with a half-period of 1600 years, and gives rise to a heavy radioactive gas, the radium emanation, now called radon. This gas is chemically inert, and in this respect belongs to the well-known group of inert gases of which helium, neon and argon are examples. The atoms of the emanation are very unstable compared with those of radium, half of them breaking up in 3·8 days. The intense radio-activity of this gas can be illustrated by a simple experiment. A minute volume of the gas, less than 1/10 cubic millimetre at ordinary pressure, is allowed to enter an exhausted glass vessel coated with phosphorescent zinc sulphide. At once the vessel is seen to glow brilliantly, due to the intense bombardment of the zinc sulphide by the vast number of α-particles shot out from the emanation as it breaks up.

It should be borne in mind that the energy liberated in the transformation of an atom, mainly in the form of kinetic energy of the α- and β-particles, is enormous compared with that released per atom in the most violent explosive. If we took a gram of pure radium salt and enclosed it in a glass tube, the α-particles shot out from radium and its products would be absorbed either in the radium salt or the glass walls, and their energy of motion ultimately converted into heat in situ. Some of the faster β-particles and most of the γ-rays would escape through the glass walls. Owing to the production of heat, it would be found that the radium tube is always a few degrees hotter than its surroundings. The emission of heat would fall off slowly with time and be reduced to one-half in 1600 years. The α-particles lose their velocity in traversing matter and ultimately lose their charge and become ordinary atoms of helium. The helium produced in this way can be released by heating or dissolving the radium salt. The enormous heat emission from a radioactive body can be best illustrated by considering a more rapidly changing product, for example, the radium emanation, which has a half-period of

3·8 days. As will be seen in Fig. 1, the emanation, after breaking up with the emission of an α-particle, gives rise to four rapidly changing products, radium A, radium B, radium C and radium C', two of which emit α-particles and two β-particles. A few hours after the emanation has been introduced into a closed tube, a kind of equilibrium is reached between the emanation and its four short-lived products, and the activity of each product then keeps step with the decay of the emanation. At the end of a month or two practically all the emanation has been transformed into radium D. This has such a long period (25 years) compared with its disintegration products, radium E and radium F, that the ultimate decay of these products is governed by that of radium D.

Suppose in imagination we could obtain a substantial quantity, say a kilogram, of radium emanation and introduce it into a heat-resisting bomb. At the end of about 2 hours, heat would be liberated at a rate corresponding to 20,000 kilowatts and the bomb would be melted unless it were cooled very efficiently. This heating effect would die away at the same rate as the decay of the emanation and would fall to half in

3·8 days. At the end of about 2 months, most of the emanation would have disappeared and the bomb would be found to contain helium gas, derived from the α-particles, equal in volume to three times that of the original emanation, while the walls of the container would be coated with a deposit of 946 grams of radium D, a slowly decaying radioactive isotope of lead of atomic weight 210. If we followed the experiment for a further 200 years, the radium D would have mostly disappeared, and we should find it replaced by an inactive isotope of lead, uranium-lead, of atomic weight 206. Owing to the liberation of α-particles from radium F, the volume of helium would have increased in the ratio of 4 to 3.

It is of interest to note that the last radioactive element in the series, radium F, generally known as polonium, was the first of the radioactive elements separated from uranium minerals by Mme Curie in 1897.

While we can predict with certainty the consequences of such an experiment as I have outlined, we are quite unable to realize it in practice, for it would require about 200 tons of radium to supply a kilogram of the emanation, while the total amount of radium so far isolated is probably under 1 kilogram. We may all be thankful too that an experiment on such a scale cannot be actually tried, for the intense emission of energy in the form of penetrating γ-rays, escaping from the bomb, equivalent to 1000 kilowatts, would certainly be dangerous to the health of those in its neighbourhood.

I hope, however, that such an imaginary experiment may serve to bring home to you the enormous emission of energy in radioactive changes, as well as the striking nature of the transformations which result in the ultimate change of the emanation into helium and uranium-lead. These radioactive transformations are spontaneous and uncontrollable. Neither intense heat nor extreme cold has the slightest effect on this natural process. We can only watch and study these wonderful changes without being in any way able to alter them.

This process of radioactivity is shown to a marked extent in the two heaviest elements, uranium and thorium, and only in a very feeble degree by a few other elements. The majority of the elements normally show no trace of radioactivity, so we may justifiably conclude that the atoms of these elements are permanently stable under ordinary conditions in our earth. During the last few years we have discovered methods not only of changing artificially one element into another, but also of producing many new radioactive elements which break up according to the same laws as the natural radioactive elements. This knowledge has only come as a result of intensive research over many years, and the development of new and powerful methods of attack on this most fundamental of problems in Physics.

ELEMENTARY PARTICLES

The transformation of the radioactive bodies brought to our knowledge the swift α- and β-particles as probable constituents of a heavy atomic nucleus. Subsequent research on the transmutation of the ordinary elements has disclosed the existence of several other types of elementary particles which are liberated as the result of explosions of atomic nuclei. The most important of these new particles are the *proton, neutron, deuteron* and *positive electron*. The proton is another name for the hydrogen nucleus of charge 1 which has a nuclear mass of 1·0076. The neutron is an uncharged particle of mass slightly greater than the proton, namely 1·0090. It is now believed that these two particles, the proton and neutron, are closely related to each other. Under the intense forces which exist in atomic nuclei, it is believed that the neutron may be converted into a proton by the removal from it of a negative electron. Conversely, the proton may be converted into a neutron by the addition of a negative electron. While we have no definite proof so far of such mutual conversions, the general evidence certainly supports the idea that there exists a definite connection between these two particles. It is natural to suppose that the neutron is a very close combination of a proton and a negative electron, although at the moment the explanation of the differences in masses between these two particles present certain difficulties.

The α-particle is a helium nucleus of charge 2, and has a nuclear mass 4·0029. The recent discovery by Urey that an isotope of hydrogen of mass 2 is always present in small proportion in ordinary hydrogen has proved of much importance to both Physics and Chemistry. By subjecting ordinary water to continued electrolysis, pure heavy water can be formed in which the hydrogen atom of mass 1 is replaced by the isotope of mass 2. This water is about 11 per cent. heavier than ordinary water and has a different freezing and boiling point. This heavy hydrogen of mass 2 has been named *deuterium* and is given the chemical symbol D. In the electric discharge through heavy hydrogen, some of the atoms lose a negative electron and become positively charged ions. These ions are called 'deuterons', while the ions of ordinary hydrogen, as we have seen, are called 'protons'. It is desirable to have distinctive names for these two ions as they are often used as fast particles to bombard matter. It will be seen that swift protons and deuterons, as well as α-particles and neutrons, have proved exceedingly effective agents in transforming many of the elements. Direct experimental evidence shows, as we should expect, that the deuteron is made up of a close union of a proton and neutron.

The positive electron, the counterpart of the negative electron of small mass, also appears in some transmutations. The discovery of this elusive particle was first made by Anderson a few years ago as a result of experiments on the cosmic rays. We are now able to produce positive electrons in small numbers in the laboratory by passing γ-rays of high quantum energy through matter. They also are liberated with high speed when certain light elements are bombarded by α-particles. The name 'positron' has been given to the positive electron, which is believed to have the same light mass as the ordinary negative electron and an identical but opposite charge.

Two other light elements or rather new isotopes of hydrogen and helium, viz. ^{3}H and ^{3}He, appear in certain transformations. Both of these isotopes appear to be stable, but neither has so far been observed in ordinary matter. It was at first thought that ^{3}H was present in preparations of heavy water, but this has not been confirmed by subsequent observations.

DETECTION OF FAST PARTICLES

We have seen that the radiations from active bodies all possess the characteristic property of discharging an electrified body. This is due to the power of the flying α- or β-particle of producing a multitude of positively and negatively charged particles, or ions as they are termed, in their passage through a gas. The primary act of ionisation is the removal from the atom or molecule of one of the outer electrons as the result of a collision with the swift particle. These ions move through the gas in an electric field, the positive ions travelling to the negative electrode and vice versa. The movement of the two kinds of ions in opposite directions is equivalent to the passage of an electric current through the gas.

In the early days of radioactivity it was customary to study and compare the effects of the radiations by this electric method, using an electroscope or electrometer as the measuring instrument. This electric method is a very convenient means of detecting minute quantities of radioactive matter and is still widely used when easily measurable effects are to be expected.

The rapid advance in recent years of our knowledge of the transmutations of the ordinary elements has been largely due to the discovery of powerful methods of detecting and counting individual atoms with high velocities such as protons, or α- or β-particles. These methods in the last resort depend on the ionisation pro-

19

duced by the flight of these particles through a gas.

The most wonderful of these methods has been devised by Professor C. T. R. Wilson and depends on the observation that the ions produced by a fast particle act as nuclei for the condensation of water vapour upon them under certain conditions. Each ion then becomes the centre of a visible drop of water. Since a fast α-particle produces more than 100,000 pairs of positive and negative ions in a gas, the actual track of the flying particle through the gas becomes visible as a crowded trail of water drops. Stereoscopic photographs of the tracks taken at the moment after expansion show clearly the position of the tracks in space.

The apparatus used for this purpose is called a 'cloud' or 'expansion' chamber, of which a typical diagram is shown in Fig. 2, with an explanation of its action. The cloud chamber is cylindrical in shape and the space A is saturated with water vapour. Suppose, for example, an α-particle passes through the gas at the moment of expansion, then if the amount of expansion is properly adjusted, each ion formed in the path of the α-particle becomes a centre of

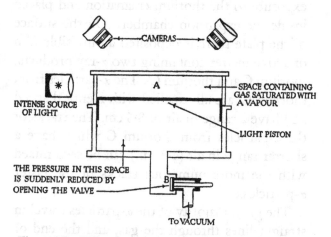

Fig. 2. The Wilson expansion chamber. The light piston is suddenly lowered by reduction of the pressure beneath it. The gas (A) above the piston expands and is cooled so that it becomes super-saturated with the vapour. This vapour condenses as small liquid drops upon any electrified particles (ions) which may be present. The chamber is illuminated through the glass wall and the condensed drops photographed by the light which they scatter into the two cameras above.

condensation, and the track of the particle is clearly seen. A photograph of the tracks of α-particles produced in this way is shown in

Plate I. The source of α-particles in this case is a small metal plate which has been activated by exposure to the thorium emanation and placed inside the expansion chamber. On the surface of the plate there is deposited an invisible film of active matter containing two α-ray products, thorium C and thorium C'. The α-particles from thorium C' are all expelled with identical speed and have a range in air of 8·6 cm. The tracks of the α-particles from thorium C which have a shorter range in air (4·8 cm.) can be seen mixed with the more numerous tracks of the faster α-particles.

The great majority of the α-particles travel in straight lines through the gas, and the end of the track represents the point where the velocity of the α-particle has fallen to such a low value that it can no longer produce ions. The passage of a β-particle through a gas gives a track which shows certain characteristic differences from that produced by the more massive α-particle. In the first place, the track is much less dense, due to the much smaller ionization per unit path produced by the β-particle. This is clearly seen in the photograph of β-ray tracks shown in Plate II. The straight track of a fast β-particle

PLATE I

(*Photographed by* PROF. J. CHADWICK.)

Tracks of α-particles emitted from thorium (C+C′) showing
the two groups of ranges 8·6 and 4·8 cm. in air.

PLATE II

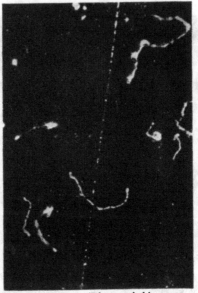

(Photographed by
PROF. C. T. R. WILSON.)

Tracks of photoelectrons of range about 1 cm. produced by the absorption in air of the characteristic K radiations from silver. (Energy about 21,000 volts.) The straight track was formed by an electron of much greater energy and was probably due to the cosmic radiation.

is marked out by a succession of drops that are so widely separated that their number can almost be counted. In the second place, the light β-particle is much more readily deflected from its rectilinear course than an α-particle of equal velocity, owing to the collision of the β-particle with atoms in its path. This is shown by the lack of straightness of the short tracks of β-particles in the photograph. The marked broadening of the end of the track is a result of the increase of ionization of the β-particle with lowering of its speed.

The photograph of the β-ray track shown in Plate III is of marked interest in showing some of the adventures of the β-particle in its passage through the gas. The long track starting on the left is sharply bent through about a right angle. This is due to the collision of the β-particle with the heavy nucleus of an atom. The short tracks diverging from the main track are due to secondary electrons which are hurled out of the atoms as the result of collisions with the fast β-particle.

The velocity and energy of a flying β-particle can be directly determined by measuring the curvature of the track in a uniform magnetic field. If the field is perpendicular to its direction of flight, the β-particle travels in a circular path. If the field is strong and the β-particle not too fast, the track of the particle produced in the gas may describe a complete circle many times in succession. The direction of deflection of the particle by the magnetic field depends on whether the charge is positive or negative. If the direction of travel of the particle is known, this method allows us to distinguish at once whether the track is due to a fast positive or negative electron.

ELECTRIC METHOD

In many experiments it is important to count the number of fast particles which enter a detecting chamber in a given time. This can be most simply done by using an electric method of counting. The way of doing this is illustrated in Fig. 3. Suppose, for example, we are counting α-particles. These are allowed to enter the detecting chamber through a thin metal foil A and are stopped by the parallel insulated plate B. A sufficient voltage is applied to remove rapidly to the electrodes the ions formed between the plates, which are usually 3 to 5 millimetres apart.

PLATE III

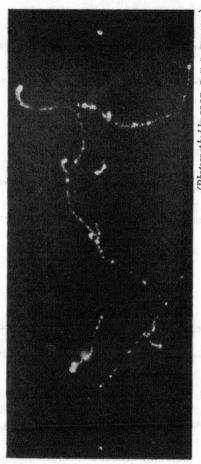

(*Photographed by* PROF. C. T. R. WILSON.)

Track of a photoelectron produced by the absorption of an X-ray quantum of energy ~ 40,000 volts. The straight initial portion of the track shows a large deflection due to a close encounter with an atomic nucleus. The increasing density of ionization and the curvature produced by collisions as the velocity diminished towards the end of the track are clearly seen.

PLATE IV

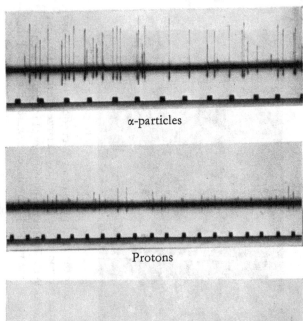

α-particles

Protons

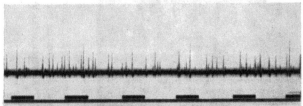

Neutron recoil particles

Oscillograph records of ionizing particles. The top two photographs were obtained under identical conditions and the difference
in the size of the deflections is due to the different ionizing
powers of α-particles and protons. The bottom photograph
shows deflections produced by helium atoms recoiling from the
impacts of neutrons of 2 million volts energy. The time intervals
are seconds.

The consequent slight rise of voltage of B, due to the entrance of a single α-particle, is automatically magnified about 100 million times by a series of amplifiers which are specially adapted for the purpose. The momentary rise of voltage on the output side is sufficiently great—of the

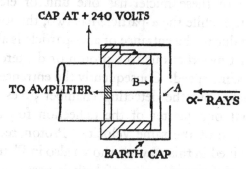

CAP AT + 240 VOLTS

TO AMPLIFIER←

B→

A

∝- RAYS

EARTH CAP

Fig. 3. Single chamber counter.

order of 100 volts—to cause a deflection of a robust form of oscillograph which has a very short period of vibration—less than 1/1000 of a second. A photographic record of the deflection of the oscillograph due to the entrance of a succession of α-particles is shown on Plate IV. The vertical lines represent the magnified electric effect of a single α-particle, while the continuous horizontal band shows the natural movement of the oscillograph while no α-particles enter. By using a rapidly moving photographic film, particles can be individually recorded which enter at a rate as fast as 1000 a minute. In a similar way, fast protons and deuterons can also be counted. Owing, however, to the fact that each of these nuclei has one unit of electric charge while the α-particle has two, the ionization due to the entrance of an α-particle is about four times that due to a proton or deuteron of the same speed. Consequently the entrance of a proton into the detecting chamber gives only about one-quarter of the deflection for an α-particle of the same velocity. Proton records obtained in this way are shown also in Plate IV. The difference in size of kick is useful in deciding whether the entering particles are singly or doubly charged, and records for protons and α-particles are easily distinguishable under certain conditions.

This method of electric counting cannot be used for fast β-particles since the ionization is too small to give a measurable deflection. A simple and sensitive method for counting β-particles has however been devised by Geiger and has come into general use. The construction of this counter is quite simple. It consists merely of an outer metal cylinder, closed at each end by an insulating cork through which passes centrally a wire or rod which is connected to a simple amplifying system. The pressure of the air or other gas in the tube is adjusted to a convenient value and a voltage is applied to the outer tube almost sufficient to cause a discharge through the gas. If a β-particle passes through the gas under these conditions, the ionization produced in the gas is greatly magnified by the well-known process of ionization by collision and a momentary discharge passes between the wire and outer cylinder. This is magnified by the amplifiers and the β-particles can be counted like α-particles either by the clicks on a telephone receiver or by an oscillograph. This Geiger-Müller tube, as it is called, is a remarkably efficient method of counting fast positive or negative electrons which enter through the cylinder walls.

Since the γ-rays give rise to β-rays in their passage through the walls and gas of the counter, this tube is also a delicate means for detecting the presence of γ-rays.

When it is required to count large numbers of fast particles, whether α- or β-particles or protons, an automatic system of counting is often used and the number of particles is recorded by a mechanical counter. Ingenious methods for this purpose have been devised by Dr Wynn-Williams and are in general use in many laboratories.

Since the neutron has no charge, it can pass freely through the outer structure of atoms without producing any ions. Occasionally, however, the neutron collides with an atomic nucleus in its path and sets it in swift motion. This recoiling nucleus is able to produce ions in a gas before it is brought to rest. These recoiling particles can be detected by the methods used for counting either α- or β-particles. It is usual for this purpose to use hydrogen, helium or air in the detecting chamber. In general not more than one neutron in 5000 which enter the counter produces a measurable kick in the oscillograph. A record of the 'neutron recoils' in helium gas is shown in Plate IV.

It will be seen later that efficient methods have been devised for counting very slow neutrons by using their property of transforming certain elements like lithium and boron.

TRANSFORMATION OF ELEMENTS
BY α-PARTICLES

When once the series of natural transformations occurring in uranium and thorium had been worked out, it was natural to hope that some day we might be able to discover methods of breaking up the stable atoms of some of the ordinary elements. Before this problem could be attacked with any hope of success, it was necessary to have some idea of the structure of the atoms of the elements. We now believe that the atoms of all the elements have a similar type of structure and are closely related to one another. At the centre of each atom is a very minute nucleus which has a resultant positive charge and which is responsible for most of the mass of the atom. The nuclear charge changes by unity in passing from one element to the next, and, as we have seen, has the value 1 for hydrogen and 92 for uranium. Surrounding the nucleus at a distance is a distribution of light negative electrons equal in number to the charge on the nucleus. The nuclear charge on an element controls the number and distribution of the outer electrons, so that the properties of an atom, as was first shown by Moseley, are defined by a whole number. Nearly every number from 1 to 92 is represented by a known element.

Some of the outer or planetary electrons of an atom can be easily removed by the electric discharge or by other methods, but after a short time other electrons take their place and the atom is reconstituted as before. If we are to effect a permanent transformation of an atom we must either alter its nuclear charge or its mass or both at once. Now the nuclei of atoms are held together by very powerful forces, and it seemed clear from the outset that very concentrated sources of energy must be brought to bear on the nucleus to effect its disruption. Twenty years ago, the most energetic particle known to Science was the fast α-particle spontaneously ejected from radium and other radioactive substances. The speed and energy of this particle is so great that it is able to penetrate deeply into the atom, and its deflection or scattering gives us important information as to the nature and intensity of the deflecting field within the atom. Indeed, the ideas of to-day of the nuclear constitution of atoms arose from a study of the scattering of α-particles through large angles as the result of their passage through matter. Suppose, for example, we consider the path of an α-particle which passes close to the nucleus of a heavy atom. Since the α-particle carries two units of positive charge and the nucleus a high positive charge, repulsive forces

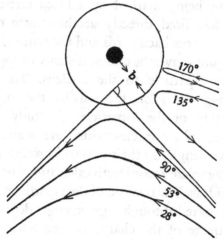

Fig. 4. Orbits of α-particles, passing close to a heavy nucleus.

come into play which become very large near the nucleus. The α-particle describes a curved path round the nucleus and, if the Coulomb law of force is obeyed, the path is an hyperbola, the asymptotes of which coincide with the direction of approach and recession of the α-particle. The α-particle may suffer a large deflection as the result of a single collision with the nucleus. The orbits of α-particles which are fired at different distances from the centre of the nucleus are illustrated in Fig. 4, the dimensions of the heavy nucleus being marked by a black circle. The α-particle fired directly at the centre of the nucleus turns backwards and the outer circle in the figure shows the closest distance of approach of the α-particle to the nucleus. The more glancing the shot, the smaller the angle of scattering of the α-particle. A study of the number of α-particles which are scattered at different angles in their passage through matter is in complete accord with calculations based on these ideas. The fraction of the α-particles which are scattered through a given angle depends on the square of the charge on the nucleus and increases rapidly with lowering of the velocity of the α-particle. It is to be borne in mind that the target area represented by a nucleus is so minute that it is only occasionally that the α-particle passes close enough to a nucleus to suffer a large deflection. An example of such a large deflection is shown in Plate V, where the α-particles are passed through oxygen. The

PLATE V

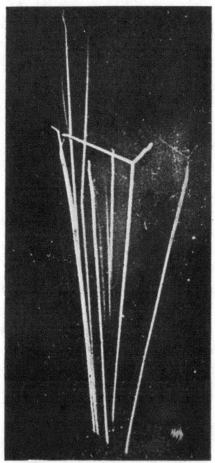

(Photographed by PROF. P. M. S. BLACKETT.)

Tracks of α-particles in oxygen, one showing a fork due to collision with an oxygen nucleus. The short branch of the fork was produced by the recoiling oxygen nucleus and the long branch by the deflected α-particle. Measurements of the angles of deflection of the two branches showed that momentum and energy were conserved in the collision.

α-particle is deflected to the left and the spur of the recoiling nucleus is shown on the right. In this discussion we are considering only 'elastic' deflections where the laws of mechanics are obeyed. In fact the colliding nuclei in this case behave like minute perfectly elastic billiard balls. No question of transmutation is involved. It is clear, however, that in a 'head-on' collision of a swift α-particle with a light nucleus of small charge, the repulsive forces are relatively smaller and may allow the α-particle to approach very close to the nucleus and possibly even to penetrate it. In such a case, the whole structure of the nucleus would be violently disturbed, possibly leading to its disintegration. Actuated by these ideas, the atoms of several light elements were bombarded by a very large number of α-particles. Under these conditions, I found evidence in 1919 that a few of the atoms of nitrogen were disintegrated with the emission of a fast hydrogen nucleus, now known as a proton. In the light of later results, the general mechanism of this transformation is well understood. Occasionally the α-particle actually enters the nitrogen nucleus and forms momentarily a new fluorine-like nucleus of mass 18 and charge 9. This nucleus, which does not exist in nature, is very unstable, and breaks up at once, hurling out a proton and leaving behind a stable oxygen nucleus of mass 17. The stages in this process of transformation are shown below in the form of a kind of chemical equation. The elements on the left represent the participants in the reaction and those on the right the final products of the transformation. The two numbers in front of each symbol represent the mass and nuclear charge of the element in question. It will be seen that nuclear charge is conserved, and so also is mass if we take into account that energy and mass are equivalent. For this purpose the symbol E on the right represents the mass equivalent of the kinetic energies of the proton and oxygen nucleus less the initial energy of the α-particle:

$$^{14}_{7}N + ^{4}_{2}He \rightarrow ^{18}_{9}F \rightarrow ^{17}_{8}O + ^{1}_{1}H + E.$$

The amount of transformation is on a small scale, for only one α-particle in about 50,000 approaches closely enough to the nucleus to be captured by it. By photographing the tracks of several hundred thousand α-particles in an expansion chamber filled with nitrogen, Blackett observed several clear cases of transformation of the nitrogen nucleus. One of these photographs

34

PLATE VI

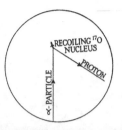

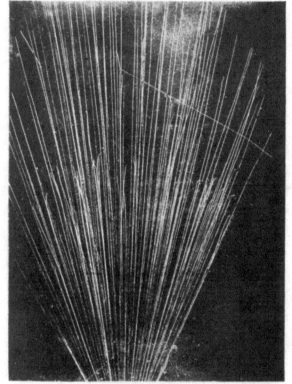

(Photographed by PROF. P. M. S. BLACKETT.)

Disintegration of nitrogen by α-particles. Of a large number of α-particles passing through nitrogen one has effected transmutation of a nitrogen nucleus into ^{17}O with the emission of a proton of high energy.

$$(^{14}_{7}N + ^{4}_{2}He \rightarrow ^{17}_{8}O + ^{1}_{1}H)$$

is shown in Plate VI and an explanatory diagram is attached. The backward track of a proton of long range and the spur of the recoiling nucleus are clearly seen.

A similar type of transformation occurs in a number of light elements exposed to α-particle bombardment, and in each case a fast proton is liberated. During the last few years the mechanism of these transformations has been closely studied and important conclusions have emerged. It is found that the protons emitted consist of two or more groups, each of definite velocity. The difference of the energy of the protons in these groups appears to be a result of emission of energy from the exploding nucleus in the form of γ-rays. There is also clear evidence of definite energy or 'resonance' levels in the nucleus leading to selective capture of α-particles of a definite speed.

DISCOVERY OF THE NEUTRON

We have seen that the proton appears as a product of the transformation of many light elements bombarded by α-particles. Still another particle of great importance has been disclosed as a result of close study of these transmutations. When the light element beryllium of mass 9 is bombarded by α-particles, no protons appear, but Bothe found that a radiation is emitted of an even more penetrating character than the most penetrating γ-rays from radium. Certain peculiarities in the absorption of this radiation were observed by M. and Mme Curie-Joliot. Chadwick finally showed in 1932 that the main radiation was not of the γ-ray type at all, but consisted of a stream of swift uncharged particles of mass about the same as that of the atom of hydrogen. These particles, which are called neutrons, have very novel properties, for, on account of the absence of charge, the neutron passes freely through the atoms in its path and produces no ionization along its track. The mechanism of the transformation which gives rise to the neutron appears to be as follows. Occasionally the α-particle is captured by the beryllium nucleus of mass 9, momentarily forming a ^{13}C nucleus with a large excess of energy. This immediately breaks up into a stable ^{12}C nucleus and a neutron, the excess of energy in the reaction appearing in the form of kinetic energy of the two final particles. A very convenient and steady source of neutrons can be made by mixing about 100 milligrams of a pure radium salt with powdered beryllium in a sealed tube. The bombardment of the α-particles produces about half a million neutrons per second, most of which pass through the containing tube. Strong sources of neutrons can also be made by using the radium emanation instead of a radium salt. In this case, the production of neutrons ultimately diminishes with time at the same rate as the emanation decays.

The possibility of the existence of neutrons as a unit in building up atomic nuclei had been envisaged long before their actual discovery. It may be of some interest to quote a statement made by the writer on this subject in the Bakerian Lecture given before the Royal Society in 1920:

If we are correct in this assumption it seems very likely that one electron can also bind two H nuclei and possibly also one H nucleus. In the one case, this entails the possible existence of an atom of mass nearly 2 carrying one charge, which is to be regarded as an isotope of hydrogen. In the other case, it involves the idea of the possible existence of an atom of mass 1 which has zero nuclear charge. Such an atomic structure seems by no means impossible. On present views, the neutral hydrogen atom is regarded as a nucleus of unit charge with an electron attached at a distance, and the spectrum of hydrogen is ascribed to the movements of this distant electron. Under some conditions, however, it may be possible for an electron to combine much more closely with the H nucleus, forming a kind of neutral doublet. Such an atom would have very novel properties. Its external field would be practically zero, except very close to the nucleus, and in consequence it should be able to move freely through matter. Its presence would probably be difficult to detect by the spectroscope, and it may be impossible to contain it in a sealed vessel. On the other hand, it should enter readily the structure of atoms, and may either unite with the nucleus or be disintegrated by its intense field, resulting possibly in the escape of a charged H atom or an electron or both.

It was thought at first that neutrons might be produced by the electric discharge through hydrogen. Experiments made with this object gave a negative result. It now seems certain that neutrons cannot be produced in this way with ordinary voltages.

Chadwick and I also made experiments many years ago to test whether neutrons were present when aluminium was bombarded by fast α-particles, but with negative results. No one could have foreseen the conditions under which this remarkable particle was ultimately discovered.

We have seen that the presence of a neutron can be detected when it makes an elastic collision with a nucleus in its path. For example, if the neutron passes through hydrogen, occasionally there is a head-on collision of the neutron with the H nucleus. The energy of the neutron is transferred to the struck nucleus, which is shot forward with a velocity equal to that of the colliding neutron. In a glancing collision, only part of the energy of the neutron is transferred. A Wilson photograph of the recoils of neutrons in methane taken by Dee and Gilbert is shown in Plate VII. By passing a stream of swift neutrons through hydrogen or material containing hydrogen, like water or paraffin, many of the neutrons are rapidly slowed down by these collisions until their energy becomes comparable with the thermal energy of the molecules surrounding them. This method of obtaining very slow neutrons has proved very useful in many experiments. Such slow neutrons pass with little absorption through large thicknesses of iron and lead, for example, but are strongly absorbed in certain elements like boron, cadmium and gadolinium. The absorption in gadolinium is so great that a layer of this material only a fraction of a millimetre thick practically removes all slow neutrons. This great absorption of slow neutrons in some elements is undoubtedly due to their capture by the nuclei concerned, resulting in a transformation of the element. Sometimes the capture of a neutron makes the resulting nucleus so unstable that it breaks up into fragments. In other cases, it may change one isotope of an element into another, one unit higher in mass, or on the other hand, it may form an unstable or radioactive isotope which breaks up with the emission of a positive or negative electron.

As Feather, Harkins and Fermi and his co-workers have shown, neutrons, and particularly slow neutrons, are extraordinarily effective agents in transforming atoms. Owing to their absence of charge, a slow neutron may freely enter a heavy nucleus, while a charged particle must have great energy of motion to get close to a heavy nucleus against the repulsion of its electric field. To illustrate the power of the neutron in transforming atoms, I will take the case of the light elements lithium and boron. A photographic method of studying neutron transformations of some elements has recently

PLATE VII

S

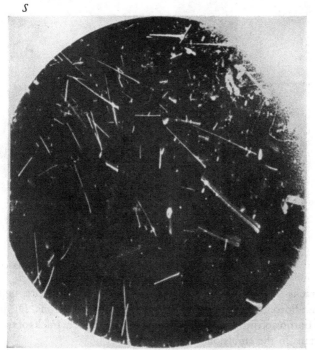

(Photographed by P. I. DEE *and* C. W. GILBERT.)

Recoil proton tracks produced in methane by the impacts of neutrons of 2·4 million volts energy. The source of neutrons was at S where a target of heavy hydrogen was bombarded by artificially accelerated deuterons.

$$({}_{1}^{2}H + {}_{1}^{2}H \rightarrow {}_{0}^{1}n + {}_{2}^{3}He)$$

PLATE VIII

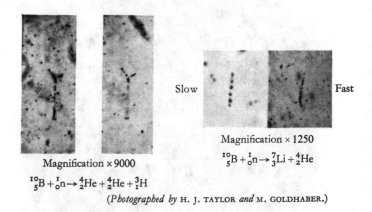

Slow   Fast

Magnification × 1250

$${}^{10}_{5}B + {}^{1}_{0}n \rightarrow {}^{7}_{3}Li + {}^{4}_{2}He$$

Magnification × 9000

$${}^{10}_{5}B + {}^{1}_{0}n \rightarrow {}^{4}_{2}He + {}^{4}_{2}He + {}^{3}_{1}H$$

(*Photographed by* H. J. TAYLOR *and* M. GOLDHABER.)

Tracks in photographic emulsion. A photographic plate, impregnated with boron, was exposed to the action of slow neutrons. Disintegration of boron according to the equations given produces tracks of developable grains in the emulsion.

been developed by Taylor and Goldhaber. A special photographic plate is impregnated with a solution of lithium or boron compound and then exposed for some days to a source of slow neutrons. On development of the plate, the tracks of fast particles are clearly visible under a high power microscope. In the case of lithium, the isotope of mass 6 captures a neutron and then breaks up into an α-particle (^{4}He) and an isotope of hydrogen of mass 3 (^{3}H). Using a high-power microscope the combined tracks of these two particles, which are shot out in opposite directions, are clearly visible on the plate. For boron, two types of transformation are observed. In one case the boron isotope of mass 10 captures a neutron and then breaks up into a nucleus of lithium of mass 7 and an α-particle (^{4}He); in the other the unstable nucleus breaks up into two α-particles and an ^{3}H nucleus. Photographs of tracks obtained in this way are shown in Plate VIII. The three tracks radiating from a point are clearly visible, the longest track being due to the singly charged ^{3}H nucleus. This marked transformation of lithium and boron by slow neutrons has proved very useful as a means of detecting and counting slow neutrons. In some cases, the detecting chamber is filled with gaseous boron fluoride; in others the walls of the chamber are lined with a boron or lithium compound.

PRODUCTION OF RADIOACTIVE BODIES

We now come to a discovery of great significance which was made by M. and Mme Curie-Joliot in 1933. They found that the bombardment of certain light elements by α-particles gives rise to radioactive elements which break up according to the same law as a natural radioactive body, emitting in the process not α- or β-particles but fast positive electrons. A single example will be taken as illustration. When boron is bombarded for some time by α-particles and then removed and tested, it is found to be radioactive, emitting a stream of positrons. The activity decays in a geometric progression with the time, falling to half-value in 11 minutes. The nature of the transformation is shown below in stages:

$$^{10}B + {}^{4}He \rightarrow {}^{14}N \rightarrow {}^{13}N + \text{neutron}.$$

Owing to excess of energy, the ^{14}N nucleus is very unstable and breaks up instantly into a more stable nucleus ^{13}N. The latter then breaks up slowly into stable ^{13}C with the emission of a positron ϵ^+, viz.

$$^{13}N \rightarrow {}^{13}C + \epsilon^+.$$

The production of this radio-nitrogen is confirmed by the fact that it can be collected as a radioactive gas which has the chemical properties of nitrogen.

It is of interest to note that the same radioactive gas can be produced by an entirely different method. If carbon is bombarded by fast protons, the following reaction occurs:

$$^{12}C + {}^{1}H \rightarrow {}^{13}N.$$

The radio-nitrogen ^{13}N so produced is identical in radioactive and chemical properties with the gas formed by the bombardment of boron with α-particles.

In a similar way, aluminium bombarded by α-particles gives rise to radioactive phosphorus of mass 30 which has a half-period of 3·2 minutes. The radio-phosphorus is converted into a stable nucleus of silicon 30 with the emission of a positron.

In the last few years, a large number of radioactive bodies have been produced not only by bombarding elements with α-particles but also by bombarding with fast protons and deuterons. Fermi and his co-workers have shown also that slow neutrons are very effective in producing radioactive bodies even from the heaviest elements. More than fifty of these radioactive bodies are now known, and in the majority of cases they break up with the emission of negative electrons (β-particles). Even the heaviest elements, uranium and thorium, are transformed by slow neutrons, and in each case give rise to a number of new radioactive bodies, but the exact interpretation of these transformations is still sub judice.

ARTIFICIAL METHODS OF TRANSFORMATION

So far we have dealt with the transformations produced by α-particles which are themselves derived from the disintegration of radioactive bodies, and by neutrons which arise from the transformation of beryllium by α-particles. The amount of radium available in our laboratories is limited, so that the transformation effects arising from the use of α-particles are in general small and can only be studied because of the exceedingly sensitive methods we have developed of counting single atoms of matter in swift motion. Ten years ago, it was recognized that far more intense streams of particles for bombarding purposes were necessary if we were to extend our knowledge of transmutation. It has long been known that the passage of an electric discharge through a gas at low pressure gives rise to a multitude of charged atoms and molecules. For example, if we cause a discharge through hydrogen, vast numbers of charged H atoms (protons) and also charged molecules are produced. The recent discovery of heavy hydrogen of mass 2, known as deuterium, has given us another projectile, the deuteron, which has proved of great importance in increasing our knowledge of transformations. A supply of protons and deuterons is readily provided by the electric discharge through hydrogen and deuterium respectively, but in order to give them a high velocity it is necessary to accelerate them in a strong electric field. This involves apparatus in some cases on an engineering scale, and voltages as high as a million volts, and, in addition, high-speed pumps to maintain a good vacuum and so prevent an electric discharge in the accelerating system. In Cambridge a high D.C. voltage has been obtained by multiplying the voltage of a transformer by a system of condensers and rectifiers. A photograph of some of the apparatus employed by Cockcroft and Walton in their pioneer experiments at Cambridge is shown in Plate IX. An illustration of the methods of obtaining and analysing streams of fast protons and deuterons for bombardment purposes is shown in Fig. 5. This apparatus was designed by Dr Oliphant and used for the study of the transmutation of light elements. We hope in the new High Tension Laboratory in Cambridge to obtain a steady accelerating voltage of 2 million volts D.C., which should give a spark about 20 feet long. In the U.S.A., Van der Graaf has devised a novel electrostatic generator to give the necessary high voltage. A machine of this kind has been used in transmutation experiments by Tuve, Hafstad and Dahl in Washington, giving steady potentials up to about 1 million volts. Professor E. Lawrence of the University of California has constructed an ingenious apparatus called a 'cyclotron' in which the charged particles are automatically accelerated in multiple stages. This involves the use of a huge electromagnet and powerful electric oscillators. A diagram of the accele-

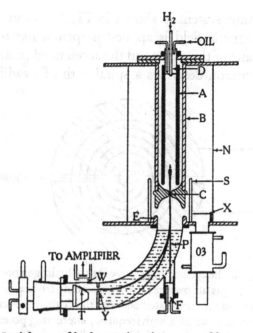

Fig. 5. A beam of hydrogen ions is generated by passing an electrical discharge through hydrogen gas at low pressure, between the oil-cooled anode A and the steel cathode B, through a hole in which at C the beam of ions issues. The ions are accelerated by potentials of up to 300,000 volts applied between the electrodes C and E, the steel shield S serving to protect the glass walls N of the apparatus. The system is exhausted by a fast pump O3 provided with a shield X. The beam of ions is passed through a magnetic field and the type of particle required is bent on to the target T through the baffle Y. The thin mica window W allows any fast particles produced to escape into the counting chamber. A Faraday cylinder F serves to collect the beam P when the magnet is not energized.

rating system is shown in Fig. 6. A uniform magnetic field is applied perpendicular to the plane of the paper and the accelerated proton or deuteron describes a spiral path of steadily in-

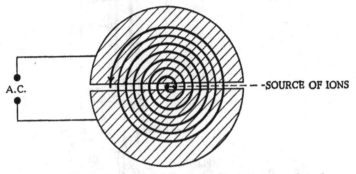

Fig. 6. The cyclotron. Positive ions of hydrogen produced in the gas at low pressure by electrons from a tungsten filament are accelerated between the D-shaped electrodes shown, to which a high frequency alternating potential is applied. A magnetic field perpendicular to the plane of the D's makes the ions move in the arc of a circle, and for a particular frequency of the applied A.C. they will always arrive at the division between the D's when the field is in the direction to accelerate them further.

creasing radius. The success of this method of multiple acceleration depends on the fact that the time required for the particle to make a complete revolution is independent of its speed,

PLATE IX

High voltage installation used at Cambridge by Cockcroft and Walton in their pioneer experiments upon artificial transmutation.

PLATE X

(*Photographed by* PROF. E. LAWRENCE.)

Six million volt beam of deuterons issuing from a cyclotron.

and consequently of the radius of revolution, provided the mass of the particle remains nearly constant. Lawrence finds that the proton or deuteron can make one thousand revolutions without serious scattering by the residual gases. In this way he has succeeded in obtaining intense streams of protons and deuterons of energies as high as 6 million volts. This energy is much greater than we can hope to obtain in the laboratory by use of direct high voltages. It is hoped to obtain still higher energy particles in the near future by using a still larger electromagnet and more intense fields.

During the past year, it has been found possible by special arrangements to allow a beam of these high-speed particles to pass outside the accelerating chamber of the cyclotron. This is a great advantage for many kinds of experiment. A photograph obtained by Lawrence of a luminous beam of deuterons corresponding to 6 microamperes and with an energy corresponding to 6 million volts is shown in Plate X. In this case the beam emerges nearly parallel through a platinum window at the end of a tube 6 feet distant from the accelerating chamber. Such a beam corresponds to the escape of

3.8×10^{13} deuterons each second, equivalent to the number of α-particles expelled each second from about 1000 grams of pure radium.

Each of these methods of obtaining fast particles has certain advantages, depending on the type of problem to be attacked.

It was at first thought that bombarding particles of the same order of energy as the α-particle, viz. about 7 million volts, would be necessary in order to penetrate the nucleus of a comparatively light element, but the development by Gamow of calculations based on wave-mechanics showed that there was a small probability of a particle entering a nucleus even though its energy was much lower than that of the α-particle. These ideas have been completely verified by later experiment. Cockcroft and Walton were the pioneers in showing that the elements lithium and boron could be transformed artificially by bombardment with protons of energy of the order of only 100,000 volts.* The modes of transformation of these elements

* This is a convenient method of expressing the kinetic energy of a fast particle. An electron, for example, in travelling in a vacuum between two points differing in potential by 1 million volts acquires energy of 1 million electron volts or, for brevity, 1 million volts.

both by protons and deuterons are now well understood and bring out many points of interest. Take first the case of lithium, which we know consists of two isotopes of masses 6 and 7. Methods have recently been devised of separating these isotopes, so that experiments can be made with targets of either ^{6}Li or ^{7}Li. Under proton bombardment, a proton occasionally enters a ^{7}Li nucleus and is captured by it. The resulting nucleus, ^{8}Be, is unstable and instantly breaks up into two fast α-particles expelled in nearly opposite directions. This type of transformation is illustrated by the diagram given in

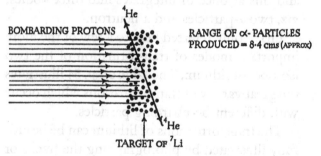

Fig. 7. Illustrating the transformation of the isotope of lithium of mass 7 by bombarding with protons. For about 100 million protons at 200,000 volts one lithium nucleus is transformed into two helium nuclei as illustrated.

Fig. 7. The capture of a proton by the ^{6}Li nucleus forms ^{7}Be, which disintegrates into an α-particle and an isotope of helium of mass 3 (^{3}He). If we use deuterons instead of protons for bombardment, the capture of a deuteron by ^{6}Li again gives rise to a ^{8}Be nucleus, but with a great excess of energy. This breaks up as before into two α-particles which have greater speed than those resulting from the capture of the proton by ^{7}Li. Indeed, with one exception, these are the fastest α-particles observed in any transformation, for they travel 13 cm. in air. The capture of a deuteron by ^{7}Li forms ^{9}Be, and this at once disintegrates into three bodies, viz. two α-particles and a neutron.

I have only referred here to a few of the more important modes of transformation of the two isotopes of lithium. The table on p. 54 illustrates the great variety of transformations which occur with different bombarding particles.

The transformations of lithium can be beautifully illustrated by photographing the tracks of the particles arising from the transformations in an expansion chamber. Such a photograph taken by Dee and Walton of the α-particles resulting from the transformation of lithium by

PLATE XI

(*Photographed by* P. I. DEE *and* E. T. S. WALTON.)

Tracks of α-particles produced in the bombardment of lithium by artificially accelerated deuterons. The α-particles which passed out to the walls of the chamber had a range > 10 cm. and were formed in the transmutation $_3^6\mathrm{Li} + _1^2\mathrm{H} \rightarrow _2^4\mathrm{He} + _2^4\mathrm{He}$ (α-particle range = 13·4 cm.). A continuous group of ranges up to 8 cm. may also be seen. This group is formed in the process $_3^7\mathrm{Li} + _1^2\mathrm{H} \rightarrow _2^4\mathrm{He} + _2^4\mathrm{He} + _0^1\mathrm{n}$. The lithium target was bombarded in vacuo within the mica window system which can be seen at the centre of the photograph below the tube down which passed the beam of swift deuterons.

PLATE XII

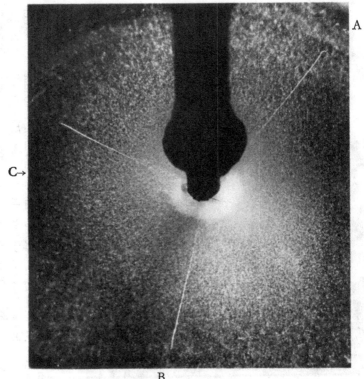

A

C→

B

(Photographed by P. I. DEE *and* C. W. GILBERT.)

The typical mode of disintegration of boron into three α-particles under proton bombardment ($^{11}_{5}B + ^{1}_{1}H \rightarrow 3\,^{4}_{2}He$). At the centre of the photograph the boron target appears as a thin line surrounded by a white sphere due to protons scattered from the bombarding beam. The α-particles A, B were emitted in nearly opposite directions while the third α-particle C received very little energy and barely emerged beyond the beam of scattered protons.

deuterons is shown in Plate XI. The appearance of pairs of particles in nearly opposite directions is clearly brought out in many photographs taken by this method.

The transformation of ^{11}B when bombarded by protons has been much studied. A ^{12}C nucleus is formed which breaks up into three α-particles. Dee and Gilbert have proved that the main type of transformation occurs in two stages. First of all an α-particle is expelled and the residual nucleus ^{8}Be is formed which contains an excess of energy, and breaks up after a very short interval into two α-particles. On account of technical difficulties, it is only occasionally that the three α-particles from a single transformation appear on the photographic plate. A fine photograph of the tracks of the three α-particles arising from a transformation is shown in Plate XII. The three particles, as they should, lie in one plane, and their total kinetic energy corresponds to the energy released in the reaction. The transformations of ^{10}B and ^{11}B when bombarded by deuterons are very complicated, groups of protons of different speeds as well as α-particles being liberated.

It would take much too long even to sum-

Transformation of lithium by bombardment with protons ($_1^1$H), neutrons ($_0^1$n), deuterons ($_1^2$H), and α-particles ($_2^4$He).

	Energy released in millions electron volts	Remarks
^{6}Li isotope:		
$_3^6$Li $+ _1^1$H $\rightarrow _2^4$He $+ _2^3$He	3·6	
$_3^6$Li $+ _0^1$n $\rightarrow _2^4$He $+ _1^3$H	4·7	
$_3^6$Li $+ _1^2$H $\rightarrow _3^7$Li $+ _1^1$H	5·0	
$\qquad\searrow$		
$\qquad\quad _2^4$He $+ _2^4$He	22	
^{7}Li isotope:		

$_3^7$Li $+ _1^1$H
$\begin{cases} _2^4\text{He} + _2^4\text{He} & \textbf{17} \\ _4^8\text{Be} + h\nu \\ \text{or } _2^4\text{He} + _2^4\text{He} + h\nu \end{cases}$ $h\nu = 17$ Homogeneous γ-ray

$\nearrow \; _2^4$He $+ _2^4$He $+ _0^1$n 14·6 Continuous distribution of neutrons

$_3^7$Li $+ _1^2$H $\rightarrow _4^8$Be $+ _0^1$n 14? Homogeneous group

$\searrow$

⟨$_3^8$Li⟩ $+ _1^1$H ? Proton group not yet observed

$_3^7$Li $+ _2^4$He $\rightarrow _5^{10}$B $+ _0^1$n − 3 Slow neutrons

In these reactions, protons, neutrons, $_1^3$H, $_2^3$He, $_2^4$He, ⟨$_3^8$Li,⟩ $_4^8$Be, $_5^{10}$B, and γ-rays are produced.

The isotope of lithium ^{8}Li marked by a circle is radioactive with a half-period of 0·8 second. It breaks up with the emission of fast β-particles. It is of interest to note that the bombardment of ^{7}Li by protons also gives rise to strong γ-rays of greater quantum energy—17 million volts—than has been hitherto detected in transformations.

marize the results obtained by bombarding all the elements by fast particles of different kinds. I would like, however, to refer to a few cases of transmutation which are of outstanding interest. The simplest possible case to consider is the effect of bombarding deuterium, ^{2}D, with

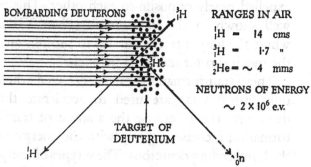

Fig. 8. Illustrating the disintegration of deuterium by bombardment with deuterons. Approximately 2 million deuterons bombarding a target of pure deuterium at 100,000 volts will produce one each of the two reactions shown.

deuterons. The coalescence of the two particles should give rise to a ^{4}He nucleus but with a very great excess of energy. This instantly breaks up in one of two distinct ways which are equally probable. This is illustrated by Fig. 8. In one case, the nucleus breaks up into a fast proton and an isotope of hydrogen of mass 3 (^{3}H), in the other into a swift neutron and a helium nucleus of mass 3 (^{3}He). If the energy of the bombarding deuteron is small compared with the energy released in the transformation, the two particles in each of the two reactions should be expelled nearly opposite to each other. This is well shown in the photograph (Plate XIII), where the longer tracks are due to protons and the much shorter tracks to ^{3}H nuclei.

These transformations can be detected when only 20,000 volts are used to accelerate the deuteron. But of course the amount of transformation increases rapidly with the energy of the bombarding deuterons. These types of transformation are the most efficient known for low-bombarding energies, and provide us for experimental purposes with a powerful homogeneous source of neutrons of energy 2·4 million volts and with a homogeneous group of protons of energy about 3 million volts.

These interesting transformations, which are of the simplest possible kind, were first studied by Oliphant and Harteck, and led to the discovery of a new isotope of hydrogen of mass 3 and a new isotope of helium, also of mass 3.

(*Photographed by* P. I. DEE.)

Three examples of the emission in opposite directions of ^{1}H and ^{3}H particles produced by the bombardment with artificially accelerated deuterons of a thin target containing deuterium ($^2_1$H + $^2_1$H → $^1_1$H + $^3_1$H). The tracks of the ^{3}H particles seen on the left of the target tube had a range of 1·6 cm. while the proton tracks emerging on the opposite side had a range of 15 cm.

The masses of these two isotopes can be deduced with accuracy from a knowledge of the energies released in the transformations. It may be of interest to show how the calculation of the mass of ^{3}H is made. If the law of conservation of energy holds, the following relation between the masses of the nuclei must hold:

$$^2H + {}^2H = {}^1H + {}^3H + E,$$
$$2 \cdot 0147 + 2 \cdot 0147 = 1 \cdot 0081 + {}^3H + \cdot 0042,$$

where E represents the mass equivalent of the energy released in the transformation. The value of E is deduced from the observed range of the protons in air, 14·70 cm., the energy of which is 2·98 million volts. Since the law of momentum must hold, it follows that three-quarters of the energy released is represented by the kinetic energy of the proton. The total energy E released is thus found to be 3·97 million volts. On Einstein's theory, mass and energy are equivalent, and a diminution of mass dm of a system is equivalent to a release of energy c^2dm, where c is the velocity of light. The correctness of this relation has been verified in a number of cases where the masses of the atoms involved in a transformation are accurately known. A release of energy of 3·97 million volts is equivalent on the atomic mass scale to ·0042. The equation given above still holds if we use the masses of the atoms on both sides of the equation instead of the masses of the nuclei. The values of the atomic masses of hydrogen and deuterium found by Aston by use of the mass spectrograph are given below the symbols. In order that the masses on both sides of the equation shall balance, it is seen that the mass ^{3}H must equal 3·0171.

In a similar way, by determining the energy of the fast neutron emitted in the alternate mode of transmutation it is found that the mass of ^{3}He = 3·0171—identical within the limits of measurement with the mass of ^{3}H. We have good reason to trust the correctness of the atomic masses deduced in this way. For example, ^{3}He appears when ^{6}Li is bombarded by protons (see table on p. 54). The mass of ^{3}He calculated from this reaction agrees with that given above.

Finally, I would like to refer briefly to some important discoveries made by Professor Lawrence and his colleagues, in which they have used the cyclotron to obtain very fast deuterons of energy as high as 6 million volts. When bismuth is bombarded by such fast deuterons, a radioactive isotope of bismuth is formed which is identical in every respect with the well-known radioactive product radium E. Not only does the radioactive body formed from bismuth emit β-particles and decay with exactly the same period as radium E, but it also gives rise to an α-ray product which is identical with polonium (radium F). The mechanism of transformation appears to be

$$^{209}_{83}Bi + ^{2}_{1}D \rightarrow ^{210}_{83}Bi + ^{1}_{1}H.$$

As shown in Fig. 1, radium E is known to be an isotope of bismuth of mass 210. This proof of the production by artificial methods of one of the natural radioactive bodies is of great interest and importance.

I must not omit to mention also another transformation which may prove to be of much technical value. When sodium of mass 23 (or common salt) is bombarded by fast deuterons, a radioactive isotope of sodium of mass 24—radio-sodium—is formed with the emission of a proton. The radio-sodium breaks up with the liberation of a β-particle forming stable nuclei of magnesium 24. The half-period of decay of radio-sodium is 15 hours. In addition to a β-particle, each nucleus of radio-sodium appears to emit a high energy γ-ray which is as penetrating as the γ-rays from radium in equilibrium with its products. Already Lawrence has succeeded by these methods in producing an intense source of radio-sodium which is equivalent in γ-ray activity to about one gram of radium. It is thus possible that such an artificially produced source of γ-rays may some day be used as a substitute for radium in therapeutic work.

TRANSFORMATION BY γ-RAYS

So far, the bombardment of matter by fast particles has proved the most effective method for studying the transformation of the elements, but in the case of the heavier elements we have seen that slow neutrons are extraordinarily efficient transformers. In some cases, however, we may expect to produce transformations by the use of γ-radiation of high quantum energy. Chadwick and Goldhaber have recently succeeded in breaking up the deuteron ^2D into a proton and neutron by the use of γ-radiation. In this case the quantum energy of the radiation must be greater than the binding energy of the proton and neutron, which is about 2·3 million volts. In a similar way, Szilard found that beryllium of mass 9 is transformed into ^{8}Be and a neutron by a γ-ray of energy not much greater than 1 million volts. This new method of transformation may prove effective in other cases if we can obtain sufficiently intense sources of γ-radiation of high energy.

GENERAL CONCLUSIONS

During the last few years, progress in our knowledge of transformation has been very rapid and almost all elements have been shown to be capable of transformation by suitable agencies. It may be of interest to record that there is good evidence that an isotope of gold can be formed by the bombardment of platinum with fast neutrons, but it is uncertain which of the isotopes of platinum is involved. In the course of this work, more than fifty new radio-active substances have been observed. These represent unstable isotopes of the elements which may have existed in the sun but disappeared as soon as the earth cooled down. It is probable that uranium and thorium are the sole survivors of a large group of radioactive elements only because their half-periods of transformation are long compared with the age of the earth. While much work remains to be done in determining the exact nature of many transformations, sufficient knowledge has accumulated to indicate that a great variety of transformations are possible with the fast projectiles available. For elements which have numerous isotopes, the possible number of transformations must be very great. In general, it is found that all possible transformations occur which are consistent with the conservation of nuclear charge and the conservation of energy when changes of mass are taken into account. The frequency of different types of transformation may, however, vary widely. In most transformations, the unstable nucleus breaks up into two particles and sometimes with the emission of γ-rays, but a few cases are known among the lighter elements in which the exploding nucleus breaks up into three particles. In the course of this work, several new stable isotopes have been detected, for example, ^{3}H, ^{3}He, ^{8}Be, ^{10}Be, which have not so far been observed in nature. In addition, as we have seen, a number of new elementary particles have been brought to light including the neutron, proton, α-particle and the positive electron.

The amount of transformation produced is

usually on a minute scale and only rarely is the quantity of matter produced either visible or weighable. Our methods of detection and recognition of the flying particles produced in a transformation, however, are so extraordinarily sensitive that even a minute amount of transformation gives very large effects in our measuring apparatus. The certainty of our methods of detection and analysis of the fast particles is in many cases greater than if the element were transformed in weighable quantities which could be analysed by ordinary chemical methods.

In general the amount of transformation produced by the bombardment method in a thick layer of the element increases rapidly as the energy of the bombarding particles is raised. In some cases no appreciable transformation is observed until the energy of the particle reaches a high value, and there is then a rapid increase of the amount of transformation with rise of energy of the particles above this threshold value. In the case of light elements like deuterium and lithium, transformation begins to be observed when the bombarding particles have energies of 20,000 volts or less. But in general the energy of the bombarding particle required to produce detectable transformation is higher and rises rapidly with the atomic number of the target element.

We have seen that these considerations do not hold in the case of neutrons, for in many cases the amount of transformation is greatest when the neutrons have low energies, of the order of a fraction of a volt, comparable with the thermal energies of the molecules. There is reason to believe also that each slow neutron which passes for example into the element boron ultimately results in the transformation of a boron nucleus (see p. 42). It seems likely too that a neutron must have a very short independent life in our atmosphere, for it would soon be captured by the nuclei of nitrogen and oxygen producing a transformation of these elements. We should consequently not expect any appreciable accumulation of neutrons in our atmosphere through the ages.

We have already discussed the large evolution of energy resulting from the spontaneous transformation of the atoms of the natural radioactive bodies. In several cases of artificial transformation by protons and deuterons, the energy emitted per disintegrating atom is even greater than from the radioactive atoms. For example, the energy release in the transformation of an atom of lithium 6 by deuterons is 22·5 million volts, nearly twice as great as the energy emitted during the disintegration of any radioactive atom. Since transformation can be produced by a deuteron of energy 20,000 volts, it is clear there is a large gain of energy in the individual process. On the other hand, only about 1 deuteron in 10^8 is effective, so that on the whole far more energy is supplied than is emitted as a result of the transformations. Even allowing for the fact that the overall efficiency of the process rises with increase of the bombarding energy, there is little hope of gaining useful energy from the atoms by such a process. At first sight the extraordinary efficiency of slow neutrons in causing transformations in certain elements with large evolution of energy seems promising in this respect, but we must bear in mind that neutrons themselves can only be supplied as the result of very inefficient processes of transformation. The outlook for gaining useful energy from the atoms by artificial processes of transformation does not look promising.

The atomic nucleus is a world of its own in which a number of particles like protons and neutrons are confined in a minute volume and held together by very powerful unknown forces. Vigorous attempts are in progress to adapt existing ideas to explain the structure of atomic nuclei and some success has been gained in a few simple cases. We are, however, still far from understanding the structure of a complex nucleus and the reason why it breaks up under certain conditions. While wave-mechanics is adequate to explain the outer electronic structure of the atom when the electrons are well separated, the theory cannot be applied with confidence to a complex nucleus when there is such an extraordinary concentration of massive particles in a very small space. To overcome these difficulties, Bohr has suggested a more general mode of attack on the problem in which the nucleus is regarded as an aggregate of indistinguishable particles which is capable of vibration as a whole and has well-marked energy levels. This new point of view has much in its favour and the prospects for the future seem more hopeful. The large amount of data now available on the transformation of the elements cannot but prove of great value in helping us to solve this most difficult and most fundamental problem.

The information we have gained on transformation processes may prove of great service too in another direction. In the interior of a hot star like our sun, where the temperature is very high, it is clear that the protons, neutrons and other light particles present must have thermal velocities sufficiently high to produce transformation in the material of the sun. Under this unceasing bombardment, there must be a continuous process of building up new atoms and of disintegrating others, and a stage at any rate of temporary equilibrium would soon be reached. From a knowledge of the abundance of the elements in our earth, we are able to form a good idea of the average constitution of the sun at the time 3000 million years ago when the earth separated from the sun. When our knowledge of transformations is more advanced, we may be able to understand the reason of the relative abundance of different elements in our earth and why, on the average, even-numbered elements are far more abundant than odd-numbered elements. We thus see how the progress of modern alchemy will not only add greatly to our knowledge of the elements, but also of their relative abundance in our universe.

Printed in the United States
By Bookmasters